Raymond Blum

Illustrated by Jeff Sinclair

GOODWILL PUBLISHING HOUSE
B-3, Rattan Jyoti, 18 Rajendra Place,
New Delhi - 110 008 (INDIA)

© Raymond Blum

This special low priced Indian reprint is published by arrangement with Sterling Publishing Co. Inc., New York, USA.

All rights reserved. No part of this publication may be reproduced; stored in a retrieval system or transmitted in any form or by any means, electronic, mechanical, photocopying or any other method, without the prior written permission of the publisher.

Published in India by:
GOODWILL PUBLISHING HOUSE®
B-3 Rattan Jyoti, 18 Rajendra Place
New Delhi–110 008 (INDIA)
Tel: 25820556, 25750801, 25755559
Fax: 91-11-25764396
E-mail: goodwillpub@vsnl.net
ylp@bol.net.in
Website: www.goodwillpublishinghouse.com

Printed at : Kumar Offset Printers, Delhi-110092

CONTENTS

A Note to Parents and Teachers   4
To Kids—Before You Begin   5

1. **Mind Reading**   7
   Skyscraper • Thumbprint • Secret Word • Multiplication Table • David Bananafield • Math Wizard

2. **Card Sorcery**   21
   Magic Spell • Whispering Deck • Lucky Joker • Easy Money • 3 Strikes You're Out • Bashful King

3. **Psychic Predictions**   37
   Mastermind • ESP • Topsy-Turvy • Perfect Match • Mind-Boggling • Millennium

4. **Paper Magic**   52
   Infinity • Magic Square • Numberland

5. **Calculator Wizardry**   59
   Jungle Math • Think of a Number • Pumpkin Pi • Spooky • Final Score • Hide-and-Seek

6. **Mixed Bag of Tricks**   74
   Tight Fit • 9-Tailed Cat • Impossible Will • Vegas • Handcuffs • Incredible Memory

Glossary   91
About the Author and the Illustrator   95
Index   96

A NOTE TO PARENTS AND TEACHERS

This book is filled with exciting and interesting mathemagic tricks that will appeal to children of all abilities, ages eight and up. Mathemagic combines magic and numbers in a way that will help spark children's interest in mathematics. All these number tricks have been classroom tested. Children love them and will enjoy performing them for family, friends, and their entire math class.

Number magic is easy for children to learn and perform because the tricks work practically by themselves. You don't need to know sleight-of-hand or have any special skills or expensive magic equipment. You can easily find most of the supplies around the house or purchase them at minimal cost.

The tricks have clear, uncomplicated, step-by-step instructions, so they are simple for children to read and understand. There is a glossary for looking up unfamiliar words. The book is organized so children can open it up, pick out a trick, and learn it on their own.

Number magic adds variety and excitement to any math class and helps make learning fun. Math teachers at any level can use this book to supplement their math program to create interest and stimulate learning. When learning is fun and exciting, children become intrigued and are motivated to learn more. This book helps provide that motivation.

TO KIDS—BEFORE YOU BEGIN

You will be able to amaze and dazzle your family and friends—and also your entire math class—with these fascinating number-magic tricks. Here are some suggestions that will help you have the most fun with this book:

1. The tricks in each chapter are organized from the easiest to the hardest. Choose the tricks that are right for you.

2. Before you perform a trick, read the directions several times, so you thoroughly understand them.

3. Practice a trick by yourself first. When you have worked it through successfully two or three times, you will be ready to perform it for others.

4. Perform each trick slowly. If you take your time, you won't make careless errors and the trick will work practically by itself.

5. Don't worry if you make a mistake. You can always blame the evil number-spirits for causing things to go wrong. Make up a magic spell that will drive them away and then try the trick again or move on to another trick.

6. Magicians never reveal their secrets. When someone asks how you did a trick, just say, "Very well!" or "It's magic!"

7. Don't repeat a trick for the same audience—they might figure out how it's done. If they want to see another trick, have a second one ready or do a variation of the same trick from the section called "Other Things to Do."

Now you are ready to perform your first magic trick for your family and friends. Good luck, be magical, and most important, have fun!

1. Mind Reading

Skyscraper

Thumbprint

Secret Word

Multiplication Table

David Bananafield

Math Wizard

SKYSCRAPER

Your friend rolls three dice when your back is turned and then stacks them on top of each other. When you turn around, you are able to reveal the sum of the five hidden faces on the dice!

What You Need

3 dice ("Monopoly," "Yahtzee," and many other board games have dice.)

What to Do

While your back is turned, have a friend:

1. Roll three dice.

2. Stack them one on top of each other.

3. Find the sum of the five hidden faces: the bottom and top of die A, the bottom and top of die B, and the bottom of die C.

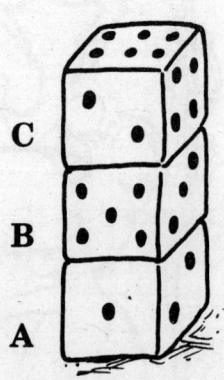

When your friend is finished, turn around and look at the number on the top face of die C.

Example: The number on the top face is 6, so mentally subtract 6 from 21 to get the sum of the five hidden faces. 21 − 6 = 15 so 15 is the secret total. Cover the stack of dice with your hands, close your eyes, and ask your friend to concentrate on the hidden sum. Pretend you are reading her mind as you reveal the secret total!

The Mathemagical Secret

On any die, the sum of the top number and the bottom number is always 7. So the sum of the top numbers and bottom numbers of three dice will always be 3 × 7 or 21.

Other Things to Do

Repeat the trick with more dice. Build your skyscraper as tall as you like. Your friend's total will equal the number of dice times 7 minus the number on the top face.

9

THUMBPRINT

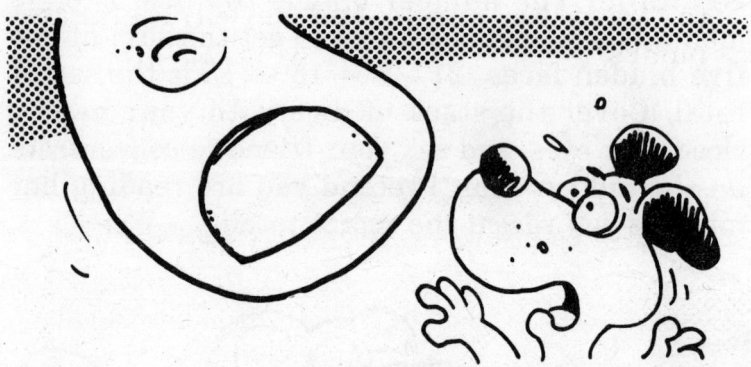

This is a trick you and your friend can perform together. After you leave the room, your friend asks someone to choose a number at random. Within seconds after you return, you are able to reveal the secret number!

What You Need

Paper and pencil

Preparation

Practice the trick with your friend before performing it for anyone else.

What to Do

1. When you leave the room, your friend asks someone to choose any number from 1 to 9. Then he writes the number on a piece of paper and folds it in half three times.

2. When you return, your friend hands you the folded-up paper. Tell him to close his eyes and concentrate on the secret number. Hold the

folded-up paper up to the side of your head, close your eyes, and pretend to be in deep thought. After a few seconds, you will be able to "read his mind" and reveal the secret number!

The Mathemagical Secret

Imagine that the top of the folded-up paper is divided into 9 equal sections. When your friend hands it to you, he signals you the secret number by simply putting his thumb over the imaginary section that contains that number.

Example: The secret number is 8.

11

Secret Word

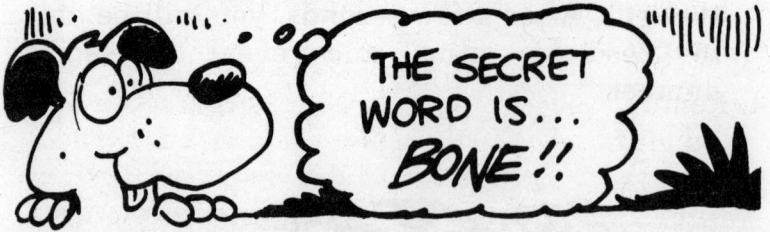

Your friend chooses a random word out of a dictionary and concentrates on it. Amazingly, you are able to read his mind and reveal the secret word in just a few seconds!

What You Need

A dictionary with more than 150 pages
A calculator   Paper and pencil

Preparation

Look at page 108 in the dictionary and memorize the 9th word down in the first column.

What to Do

1. Ask your friend to write any 3-digit number on a piece of paper without letting you see it. Tell him the first digit must be *at least 2 greater* than the last digit.

    Example
    752

2. Ask him to reverse the three digits and write this new number (257) below the first number. Tell him to subtract the two numbers on a calculator.

    752
    −257
    ‾‾‾‾
    495

12

3. Tell him to reverse the difference and add this new number (594) to the calculator total.

$$\begin{array}{r} 495 \\ +594 \\ \hline 1{,}089 \end{array}$$

4. Remind your friend he was free to choose any 3-digit number. Then when your back is turned, ask him to look at the *first 3 digits* of his final total and turn to that page in the dictionary. Next, tell him to look at the *last digit* of his final total and *carefully* count that many words down in the first column. It will be the word that you memorized.

page→<u>1 0 8</u>   <u>9</u>←words down

Finally, ask him to concentrate on that word for about 10 seconds. You should have no problem "reading his mind" and revealing the secret word!

The Mathemagical Secret

No matter which 3-digit number your friend starts with, the final total will always be 1,089.

MULTIPLICATION TABLE

This is another trick you and your friend can perform together. After your friend leaves the room, you ask someone to choose two numbers from 1 to 9 and multiply them together. When your friend returns, not a single word is spoken, yet she is able to reveal the answer!

What You Need

A table   Paper and pencil

Preparation

Practice the trick with your friend before performing it for anyone else.

What to Do

1. After your friend leaves the room, ask someone to choose any two numbers from 1 to 9 and then multiply them together.

   Example: 5 and 7

   $5 \times 7 = 35$, so 35 is the secret number.

2. Write the secret number on a piece of paper, fold it in half three times, and then put the paper and pencil on the table.

3. When your friend returns, close your eyes and pretend to be in deep thought. Your friend will have no problem "reading your mind" and revealing the secret number!

The Mathemagical Secret

Mentally divide the top of the table into 9 equal sections. You signal the secret number to your friend by placing the paper and pencil inside two of those sections. Put the folded-up piece of paper in the tens-digit section of the table (the 3) and lay the pencil in the ones-digit section (the 5).

Secret Number = 35

1	2	3
4	5	6
7	8	9

Exceptions

1. If the secret number is a single digit, don't lay the folded-up piece of paper on the table.
2. If the secret number ends in 0 (10, 20, 30, or 40), don't lay the pencil on the table.

15

DAVID BANANAFIELD

The star of this trick is a talented and magical piece of fruit—the amazing Mr. David Bananafield!

What You Need

A calculator
A thick needle
1 banana

Cellophane tape
A pen
1 facial tissue

Preparation

You might need an adult to help with the first 3 steps.

1. Carefully push the point of the needle into the center of the banana at one of its edges (Figure A).

2. Move the needle left and right until the banana has been

A

sliced in half. Be careful not to cut the skin with the point of the needle.

3. Gently pull out the needle and repeat Step 1 and Step 2 above and below your first cut. Now your banana is cut into 4 pieces (Figure B).

B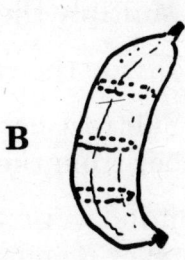

4. Carefully squeeze the banana a little near each of the 3 needle holes to close them.

5. Draw a little face on the top of the banana. Tape a facial tissue around it so it looks as if it were wearing a magician's cape, and you have created Mr. David Bananafield (Figure C).

C

What to Do

Introduce David Bananafield to your friend and explain that he is a master magician and a world famous mind reader. Then have your friend:

1. Enter any number that is easy to remember into a calculator without letting Mr. Bananafield see it. (This number must be less than 8 digits.)

   Example
   99

2. Multiply this number by 4. $99 \times 4 = 396$

3. Add 16 to the result. $396 + 16 = 412$

17

4. Multiply the answer by 3. $412 \times 3 = 1{,}236$

5. Divide the result by 12. $1{,}236 \div 12 = 103$

6. Subtract her original number from the total. $103 - 99 = 4$

Remind your friend she was free to choose any number. Then ask her to look at the calculator and concentrate on her final total. Put David Bananafield near her head so he can read her mind more easily. Finally, remove his little cape, carefully peel him, and show your friend that Mr. Bananafield has revealed her secret number by slicing himself into 4 pieces!

The Mathemagical Secret

This trick was written using a kind of mathematics called algebra. Multiplying by 4 and then by 3 is the same as multiplying by 12. Dividing by 12 cancels those operations. Every other operation is just mathematical hocus-pocus that eliminates your friend's original number and guarantees that the final total will always be 4.

MATH WIZARD

You ask your friend to choose a number from 1 to 99. Then you phone a person who lives in your area who is known as the Math Wizard. When you hand the phone to your friend, the Math Wizard reveals her secret number!

What You Need

A telephone   Paper and pencil

Preparation

The "Math Wizard" is another friend of yours. Practice the trick with him before performing it for anyone else. Also, let him know when you are going to call.

What to Do

1. Tell your friend that an amazing person called the Math Wizard lives nearby. The Wizard is an ESP expert who can read people's minds over the phone. No one knows who this person is, but it is known that he has astounding mental powers.

2. Ask your friend to think of any number from 1 to 99 and write it on a piece of paper. Sneak a peek at her secret number.

Example: 58

3. Call the Math Wizard. Don't let your friend see the phone number or she might recognize it. Say it is an unlisted phone number and you have been told not to reveal it to anyone.

4. When the Wizard answers say, "Is the Wizard there?" This is the cue for him to *slowly* and *quietly* start counting, "0, 1, 2, 3, 4, 5..."

5. When he says the tens digit of the secret number, stop him by saying, "Could I speak to the Wizard?" This is the cue for him to *slowly* and *quietly* start counting, "0, 1, 2, 3, 4, 5, 6, 7, 8..."

6. When he says the ones digit of the secret number, stop him by saying, "Hello Wizard, would you please tell my friend her secret number?"

7. Hand your friend the phone and when she says, "Hello," the Math Wizard says "58" in a disguised voice and immediately hangs up!

An Exception

When your friend chooses a 1-digit number, the tens digit is 0.

Example: 5 = 05

2. Card Sorcery

Magic Spell

Whispering Deck

Lucky Joker

Easy Money

3 Strikes You're Out

Bashful King

Magic Spell

Your friend secretly chooses a card from a deck of cards. When a magical phrase is spelled out, your friend's secret card mysteriously appears!

What You Need

A deck of playing cards

What to Do

1. Cut the deck into seven piles and place them facedown on the table.

2. Ask your friend to point to one of the piles. (It doesn't matter which one she chooses.) Gather together all the piles she did not choose into one pile and then put her chosen pile on the top.

3. Tell her to secretly look at the top card, memorize it, and return it to the top of the deck.

4. Ask her for any number between 20 and 30. (20 and 30 are not between.)

Example: 23

5. Deal that number of cards into a small pile, one card at a time. Place the rest of the deck next to the small pile.

6. Ask your friend to find the sum of the digits of her number.

$$23 \rightarrow 2 + 3 = 5$$

7. Return that many cards to the top of the big pile, one card at a time.

8. Put the small pile on top of the big pile.

Finally, hand your friend the deck. As you slowly spell the magical phrase "h-o-c-u-s p-o-c-u-s a-l-a-c-a-z-a-m," tell her to turn over one card for each letter. When she turns over the *last card*, it will be her secret card!

The Mathemagical Secret

Any number between 20 and 30 minus the sum of its digits always equals 18. "Hocus pocus alacazam" has exactly 18 letters.

WHISPERING DECK

It is easy to predict your friend's secret card when you have a deck of cards that talks!

What You Need

A deck of playing cards—exactly 52 cards

What to Do

1. Have your friend shuffle the deck of cards as many times as he wants and then hand you the deck.

2. Count out exactly 26 cards from the top of the deck. Deal them *faceup* into a pile, one card at a time. While you are counting, *memorize the 10th card*. This will be your friend's secret card.

3. Turn that pile over and put it on the *bottom* of the deck.

4. Deal out six cards facedown from the top of the deck and spread them on the table. Ask your friend to turn any three of these cards faceup.

5. Put the three cards that were not chosen on the *bottom* of the deck and leave the other three cards faceup on the table.

EXAMPLE

6. Hand your friend the deck and tell him to deal cards facedown below each of these cards. He should start with the number on the faceup card (all face cards = 10 and Aces = 1) and then keep dealing cards until he gets to 10. For example, if the faceup card is a 6, he would deal four more cards to get to 10.

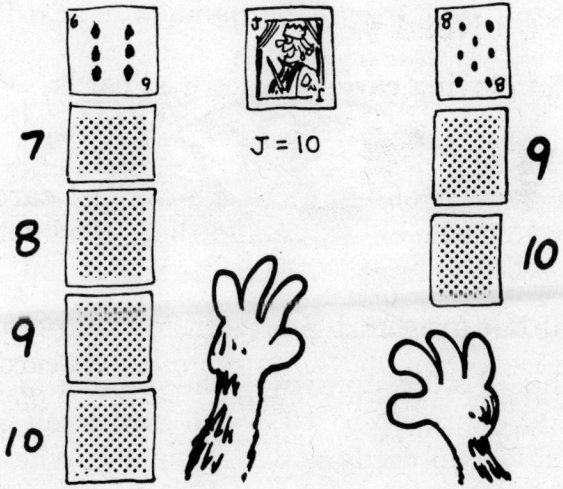

7. Tell him to keep the three faceup cards on the table and then put all the facedown cards on the *bottom* of the deck.

25

8. Ask him to find the sum of the three faceup cards.

Tell him to count that many cards down in the deck and look at the *last* card without showing you. It will be the card that you memorized.

9. Explain to your friend that you have a talking deck and that it will whisper the name of his secret card in your ear. Hold the deck up to your ear, pretend that it is whispering to you, and then tell your friend the name of his card!

The Mathemagical Secret

The value of a faceup card + the number of cards needed to count up to 10 + the card itself = 11. So three faceup cards = 3 × 11 or 33. When you add the 3 cards that were not chosen you get 36. In Step 3, the memorized card became the 36th card down from the top of the deck.

LUCKY JOKER

You are able to remove odd and even cards from the deck just by spelling! However, the trick only works if you insert the Joker into the deck for good luck!

What You Need

A deck of playing cards with a Joker

Preparation

1. Remove any 13 cards from the deck. Seven of the cards must be odd numbers and six of them must be even numbers (Jack = 11, Queen = 12, and King = 13). Remove the Joker and put it aside until later.

2. Arrange the 13 cards in a pile in the following order. (E = even and O = odd)

    (bottom) E-E-E-O-E-O-O-E-O-E-O-O-O (top)

3. Put this 13-card pile on top of the rest of the deck.

What to Do

1. Count out 13 cards from the top of the deck into a small pile, one at a time. This will reverse the order of the 13 cards you arranged earlier, so they are now in the correct positions. Set the rest of the deck aside.

2. Show your friend the pack of cards. Say you are going to make odd-numbered cards and even-numbered cards appear magically by just spelling "odd" or "even."

3. Explain that the trick only works when a Joker is inserted into the pack for good luck. Fan the pack of cards facedown in your hand and insert the Joker into the *sixth* position down from the top. Pretend that you are randomly inserting the card into the pack.

    (bottom) O-O-O-E-O-E-O-O-JOKER-E-O-E-E-E (top)

4. Hold the pack of cards facedown in your hand. Say "O" as you put the top card *at the bottom* of the pack. Say "D" as you put the next top card at the bottom of the pack. Say "D" as you put one more card at the bottom of the pack. Then say "Odd" as you turn over the *next* card. It will be odd. Remove that card from the pack and place it faceup on the table.

5. Use the same method to remove an even card. Then *alternate* removing odd cards and even cards. Remember, when you say the last letter of a word, put a card *at the bottom* of the pack and then turn over the *next* card. It will be the kind of card that you just spelled.

At the end of the trick, there will be only one card remaining. It will be the card that guarantees this magic trick will work—the lucky Joker!

The Mathemagical Secret

Place 14 cards *faceup* on the table in this order, and you will see why this trick works.

E-E-E-O-E-JOKER-O-O-E-O-E-O-O-O

Point to the first card and say "O," the second and say "D," and the third and say "D." Then say "Odd" as you point to the *next* card. It will be odd. Remove that card from the table. Continue moving to the right and use the same method to remove an even card. Alternate spelling odd and even and always remove the *next* card. When you get to the end of the row, go back to the beginning of the row and continue spelling. In the end, only the Joker will remain.

Easy Money

Your friend selects a card and then shuffles it back into the deck. When you attempt to find his card, it appears that you have made a mistake. However, it is your friend who makes a mistake—by doubting your magical powers!

What You Need

A deck of playing cards with no Jokers

Preparation

Put your favorite card nine cards up from the *bottom* of the deck.

What to Do

1. Place the deck of cards on the table. Tell your friend to cut the deck in half as best he can by taking half of the cards off the top and putting them in a pile facedown on the table.

2. Tell him you will count the remaining cards to see how close he came to cutting the deck exactly in half (26 cards). Pick up the *bottom*

half of the deck and count the cards one at a time into a small pile. Then put that pile on top of his pile. It does not matter if the deck has been cut exactly in half. This is just a little hocus-pocus to get your favorite card nine cards down from the top of the deck. If the pile does not have exactly 26 cards just say, "Close enough."

3. Ask your friend for a number *between* 10 and 20 (10 and 20 are not between).

Example: 17

4. Deal that number of cards into a small pile, one card at a time. Place the rest of the deck next to the small pile.

5. Ask your friend to find the sum of the digits of his number.

$$17 \rightarrow 1 + 7 = 8$$

6. Return that many cards to the top of the big pile, one card at a time.

7. Put the small pile on top of the big pile. Now the top card will be your favorite card. Tell your friend to secretly look at it, memorize it, and return it to the center of the deck. Tell him to shuffle the cards as many times as he wants and then hand them to you.

8. Explain to your friend that you will reveal his secret card by dealing the cards faceup on the table. Tell him not to say anything if you make a mistake and pass by his card.

9. Pretend to be counting cards as you deal them one at a time on the table. Keep going until you are *well past* your friend's secret card.

10. Suddenly stop your count and say, "I bet you one million dollars the next card I turn over will be your card." Of course, he will think that you mean the next card on the top of the deck in your hand. When your friend agrees to bet, reach into the cards that you have already dealt on the table and find his secret card. Pick it up, turn it facedown, and collect one million dollars!

The Mathemagical Secret

Any number between 10 and 20 minus the sum of its digits always equals 9.

3 Strikes You're Out!

You have only three chances to find a chosen card in the deck. Your first two tries are incorrect, but don't worry, you won't strike out. You'll find the secret card on your third try every time!

What You Need

A deck of playing cards with a Joker

What to Do

1. Have your friend shuffle the deck as many times as she wants.

2. When she hands you the cards, say you forgot to take out the Joker. Turn the cards over, remove the Joker, and sneak a peek at the top card in the deck. This is the secret card.

3. Tell your friend that magicians consider the (name the secret card) to be one of the most difficult cards to locate, but with her help, you will force it to magically appear.

4. First, say that you will look for the secret card in the top half of the deck. Ask your friend for any number between 1 and 26.

Example: 14

Deal out 14 cards facedown into a pile on the table, one card at a time. Then turn over the *next* card in your hand. It won't be the secret card, so return it facedown to the top of the pile in your hand. Then put the pile that is on the table on top of the pile that is in your hand.

5. Second, say that you will look for the secret card in the bottom half of the deck. Ask your friend for any number between 26 and 52.

Example: 35

Deal out 35 cards exactly as you did in Step 4 and turn over the *next* card in your hand. It won't be the secret card either, so return it face-down to the top of the pile in your hand. Then put the pile that is on the table on top of the pile that is in your hand.

6. Finally, ask your friend to subtract her two numbers.

$$35 - 14 = 21$$

Count that many cards down in the deck and then turn over the *next* card. This time it will be the secret card!

The Mathemagical Secret

When you count the cards twice, the top card moves to a different place in the deck. Simple subtraction tells you the number of places that it moves down from the top of the deck.

Bashful King

Your friend chooses a King from a deck of cards. It turns out the King is very shy, however, and refuses to show his face until the very end of the trick!

What You Need

A deck of playing cards—exactly 52 cards

Preparation

1. Remove the 4 Kings and split the rest of the deck into two piles. The first pile must have *exactly* 27 cards and the second pile *exactly* 21 cards.

2. Place the 4 Kings faceup on the table. Put the 27-card pile facedown on the left side of the Kings and put the 21-card pile facedown on the right side. The two piles are almost the same height, so it looks as if the deck has been cut in half.

What to Do

1. Ask your friend to pick up his favorite King. Then shuffle the other three Kings into the 27 card pile.

2. Tell your friend to put his King facedown on top of that pile.

3. Place the 21-card pile on top of his King so you now have a 52-card deck with your friend's King somewhere in the center.

4. Explain to your friend that he chose the most bashful King in the deck. His King is so shy that

35

he won't show his face again until the end of the trick.

5. Take the deck in your hand and deal the top card faceup on the table in front of your friend. Then deal the next card facedown on the table in front of you. Alternate dealing cards faceup in front of your friend and facedown in front of you until all the cards have been dealt into two piles. There will be no sign of your friend's shy King. Put the faceup pile aside.

6. Pick up the facedown pile and deal it faceup and facedown as you did in Step 5. When you are finished, put the faceup pile aside.

7. Repeat Step 6 three more times and you will end up with only one facedown card in front of you. It will be your friend's shy King!

The Mathemagical Secret

This trick works because of odds and evens. Each card that is an odd number down from the top of the deck is turned up, and each card that is an even number down from the top of the deck is turned down. When you put the 21-card pile on top of your friend's chosen King in Step 3, it becomes the 22nd card down from the top of the deck—an even number. Then as the trick progresses, his chosen King becomes the 16th card, 6th card, 4th card, and the 2nd card down from the top of the deck—all even numbers.

3. Psychic Predictions

Mastermind

E.S.P.

Topsy-Turvy

Perfect Match

Mind-Boggling

Millennium

MASTERMIND

Normally, there would be only a one-out-of-three chance that you would correctly predict the outcome of this trick. However, when you know the mathemagical secret, you will be correct every time!

What You Need

A deck of playing cards   Paper and pencil

Preparation

Put three small piles of cards facedown on the table. The first pile is made up of four 7's, the middle pile has any seven cards in it, and the third pile has four Aces and a 3-card.

What to Do

1. Tell your friend you are going to make a prediction. Then secretly write this on a piece of paper:

    You will choose the 7 pile!

Fold your prediction several times and put it aside until later.

2. Ask your friend to point to one of the piles.

If she points to the first pile (four 7's), gather up the other two piles and shuffle them back into the deck as your friend turns over her pile—which has the four 7's in it.

If she points to the middle pile (any 7 cards), pick up the other two piles and count them face-down, one card at a time, to show one contains 4 cards and the other contains 5 cards. Then shuffle them back into the deck as your friend counts her pile—which has 7 cards in it.

If she points to the third pile (four Aces and one 3-card), gather up the other two piles and shuffle them back into the deck, as your friend turns over the cards in her pile and finds that their sum equals 7.

Finally, remind your friend she was free to choose any of the 3 piles. Then hand her your prediction and tell her to unfold it. She will be amazed to discover that your prediction was correct. She did choose the 7 pile!

ESP

Your friends will think you have extrasensory perception when you predict the final answer to a problem before any numbers are chosen!

What You Need

Paper and pencil   A calculator   A ruler

Preparation

Carefully copy ESP Chart I onto a piece of paper.

ESP Chart I

1	2	3	4
5	6	7	8
9	10	11	12
13	14	15	16

What to Do

1. Tell your friend you're going to make a prediction. Then secretly write this on a piece of paper:

   The sum of the 4 numbers you choose will be 34!

   Fold your prediction several times and put it aside until later.

2. Hand your friend ESP Chart I and tell him to circle any number. Then tell him to cross off all other numbers in that same row and column. So he would cross off all the numbers to the left and right of his circled number and all numbers above and below it.

40

Example: He circles the 6.

1	2̷	3	4
5̷	⑥	7̷	8̷
9	10̷	11	12
13	14̷	15	16

3. Ask him to circle any other number that isn't already circled or crossed off. Tell him to cross off all the numbers in that same row and column as he did in Step 2.

Example: He circles the 4.

1̷	2̷	3̷	④
5̷	⑥	7̷	8̷
9	10̷	11	12̷
13	14̷	15	16̷

4. Tell him to follow these same directions until four numbers have been circled and all the other numbers have been crossed off. After he circles the 4th number, there should be no numbers left to cross off.

Example: He circles the 9 and then the 15.

1̷	2̷	3̷	④
5̷	⑥	7̷	8̷
⑨	10̷	11̷	12̷
13̷	14̷	⑮	16̷

5. Have your friend find the sum of the four numbers he circled and announce the total. This total will always be 34.

Finally, unfold your prediction and show your friend it matches his total!

The Mathemagical Secret

In any square array of consecutive numbers, the sum of the circled numbers equals the sum of the numbers in either diagonal. ESP Chart I is a 4 × 4 array, and the sum of each diagonal is 34.

Other Things to Do

1. If your friend wants you to repeat the trick, use ESP Chart II. If you follow the same procedure and have your friend circle five numbers, the sum will always be 65. This trick works because the sum of each diagonal is 65.

ESP Chart II

1	2	3	4	5
6	7	8	9	10
11	12	13	14	15
16	17	18	19	20
21	22	23	24	25

2. Make larger charts of consecutive numbers. The sum of the circled numbers will be the sum of the numbers in either diagonal.

TOPSY-TURVY

You will amaze your friends when a randomly chosen number matches the sum of three cards that are upside down inside a sealed box!

What You Need

A deck of playing cards and its box
Paper and pencil   A calculator (optional)

Preparation

Remove any three cards from a deck of cards whose sum = 18 (example: 3 of Hearts, 9 of Clubs, and 6 of Diamonds). Put the three cards *upside down* in the deck, put the deck back inside its box, and then close the lid.

What to Do

Hand your friend the box of cards and ask her to keep it until the end of the trick. Then have her:

1. Write down any 3-digit number on a piece of paper without letting you see it. Tell her all 3 digits must be different.

   Example
   358

2. Reverse her number and write it next to her first number. 358 853

3. Subtract the smaller number from the larger number.

$$\begin{array}{r}853\\-358\\\hline 495\end{array}$$

4. Mentally find the sum of the digits of her answer. $4 + 9 + 5 = 18$

Finally, remind your friend that she was free to choose any 3-digit number and that the box of cards has not been touched since the trick started. Ask her for her final total and then tell her to open the box of cards. Ask her to spread the cards faceup on the table and notice three cards are upside down. Have her turn them over and find their sum. Your friend will be amazed that the sum of the three cards matches her final total!

The Mathemagical Secret

If you take any 3-digit number whose digits are all different, reverse the number, and then subtract the smaller number from the larger, the difference will always be one of the following answers: 99, 198, 297, 396, 495, 594, 693, 792, or 891. The sum of its digits will always be 18.

PERFECT MATCH

Your friend won't believe his eyes when his secret number matches your sealed prediction!

What You Need

Paper and pencil   An envelope

Preparation

1. Secretly write this prediction on a piece of paper:

    You will choose the number 13!

2. Fold it, seal it inside an envelope, and set it aside until later.

3. *Carefully* copy the Psychic Chart onto another piece of paper.

Psychic Chart

19	6	14	9	15
8	10	18	3	7
12	4	13	20	2
5	17	1	16	11

What to Do

Hand your friend the Psychic Chart and a pencil. Tell him he will be choosing number at random by moving around the array of numbers. Explain that he can move horizontally, vertically, forward, or backward, but *never diagonally*. Tell him he can not move onto numbers that have been crossed off. Remind him to follow the directions carefully and always keep one finger on the array even when he crosses off numbers.

Tell your friend you have predicted the number he will choose and you have sealed your prediction inside an envelope. Then have him:

1. Place his finger on any odd number.

2. Move up or down to the *nearest* even number.

3. Move left or right to the *nearest* odd number, and then cross off 19, 8, 12, and 5.

4. Move down or right one number, and then cross off 6, 14, 9, and 15.

5. Move up or left one number, and then cross off 7, 2, 11, and 10.

6. Move two numbers in any direction (not diagonally), and then cross off 4 and 18.

7. Move three numbers in any direction, and then cross off 3 and 17.

8. Move four numbers in any direction, and then cross off 16.

9. Move one more number in any direction and keep his finger there.

If your friend followed your directions, his finger will be on the number 13. 13 will always be the selected number. Remind your friend he was free to start wherever he wanted and free to move wherever he wanted every step of the way. Then hand him the envelope and ask him to open your prediction. Of course his chosen number and your prediction will be a perfect match!

The Mathemagical Secret

The array is designed so it is only possible to land on certain numbers. Those numbers that are impossible to land on are removed, and in the end, your friend is forced to choose the number 13.

MIND-BOGGLING

Is it possible to predict the answer to an addition problem before any numbers are written down? It is—if you know the mathemagical secret!

What You Need

A calculator   Paper and pencil

What to Do

1. Tell your friend you are going to make a prediction. Then secretly write the number **1998** on a piece of paper, fold it several times, and set it aside until later.

2. Ask your friend to write a 3-digit number on a piece of paper. The digits must be different and not form a pattern.

    Example

    379 ⟶ sum = 999

3. Tell her to write a second 3-digit number below the first number.

    852

4. Explain that since she wrote two numbers, you will also write two numbers. First, write a 3-digit number so that the sum of the first and third numbers = 999.

    620 ⟶ sum = 999

5. Then write a 3-digit number so that the sum of the second and fourth numbers = 999.

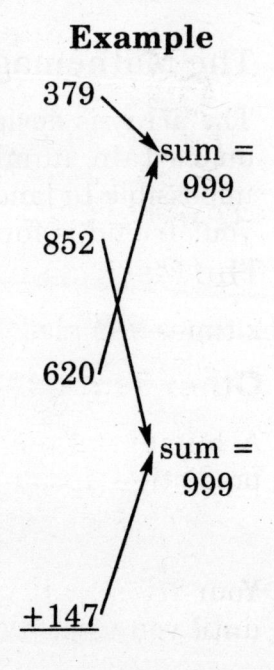

    +147

48

6. Ask your friend to add the four numbers using a calculator. The answer will always be 1998.

Finally, unfold your prediction and show your friend. She won't believe her eyes. It is absolutely mind-boggling, because it matches her final answer!

An Exception

If your friend writes a number in the 900's, you will have to write a 2-digit number.

Example:

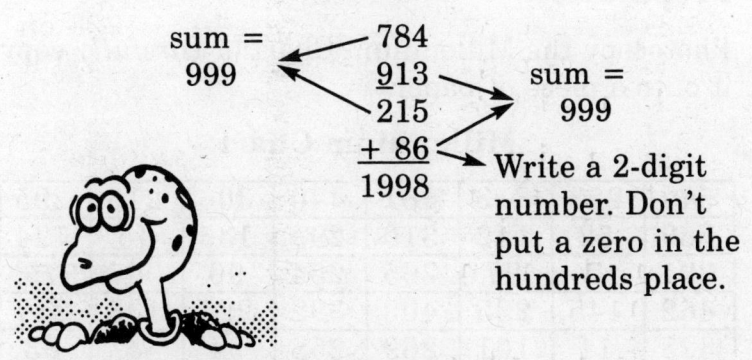

```
sum =        784
 999         913        sum =
             215         999
            + 86
            1998    Write a 2-digit
                    number. Don't
                    put a zero in the
                    hundreds place.
```

The Mathemagical Secret

2 times 999 always equals 1998.

Other Things to Do

A humorous end to this trick is opening your prediction upside down:

8661

Your friend will think you have made a mistake until you turn the paper over and it reads:

1998

MILLENNIUM

You secretly make a prediction before the trick begins. Then a friend chooses eight numbers at random and adds them together. Your prediction is unfolded and it matches his answer!

What You Need

Paper and pencil   A ruler   A calculator

Preparation

Photocopy the Millennium Chart or *carefully* copy it onto a piece of paper.

Millennium Chart

540	223	313	481	470	306	216	295
369	52	142	310	299	135	45	124
324	7	97	265	254	90	0	79
462	145	235	403	392	228	138	217
328	11	101	269	258	94	4	83
667	350	440	608	597	433	343	422
326	9	99	267	256	92	2	81
460	143	233	401	390	226	136	215

What to Do

1. Tell your friend you are going to make a prediction. Then secretly write this on a piece of paper:

    The sum of the 8 numbers you choose will be 2000!

2. Hand your friend the Millennium Chart. Ask him to circle any number in the chart. Then tell

him to cross off all other numbers in that same row and column. So he would cross off all numbers to the left and right of his circled number and all numbers above and below his circled number. (See ESP for an example.)

3. Ask him to circle any other number that isn't already circled or crossed off. Tell him to cross off all numbers in that same row and column like he did in Step 2.

4. Tell him to follow these same directions until eight numbers have been circled and all other numbers have been crossed off. After he circles the 8th number, no numbers will remain.

5. Have your friend find the sum of the eight numbers he circled on a calculator and announce the total. It will always will always be 2000.

Finally, unfold your prediction and show your friend it matches his total!

The Mathemagical Secret

The sum of each diagonal is 2000.
Then the same numbers are added or subtracted in each row and column to complete the rest of the chart.

Other Things to Do

For 2001, add 1 to each of the eight numbers in the first row. For 2002, add 2, and so on. Then, not only will your friend's total match your prediction, it will also match the current year!

4. Paper Magic

Infinity

Magic Square

Numberland

INFINITY

A piece of paper has exactly two sides—a front side and a back side. It's impossible to create a piece of paper that has only one side. Or is it? Actually, it's easy when you know the mathemagical secret!

What You Need

An 8½ in × 11 in (21.5cm × 28cm) sheet of plain paper

A pair of scissors
A pencil
Cellophane tape

Preparation

1. Cut a strip of paper that is about one inch (2.5cm) wide from the longest edge of a plain piece of paper.

2. Give the strip a *half twist* and then tape the two ends together.

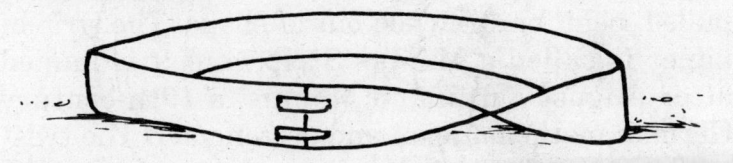

What to Do

1. Hold the strip of paper in one hand and start drawing a line straight down the middle.

2. Continue drawing the line until you get back to where you started.

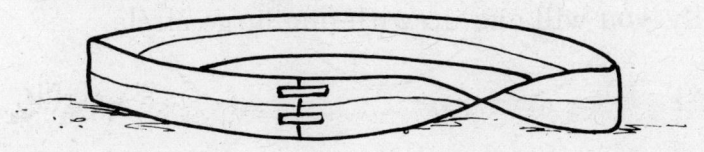

Now look at the strip of paper. Notice you drew one continuous line all the way around the paper, and you never crossed over an edge. On a piece of paper that has two sides, you would have to cross over an edge to draw one continuous line from one side to the other. Therefore, this strip of paper only has only one side!

The Mathemagical Secret

This trick uses a kind of mathematics called topology. Topology is the study of shapes and what happens to those shapes when they are folded, pulled, bent, or stretched out of shape. The strip of paper is called a Möbius Strip, and it is named after August Ferdinand Möbius, a 19th-century German mathematician and astronomer. The twist in the strip of paper connects what was once the front side to the back side so there is only one big surface.

Other Things to Do

Cut the strip of paper along your pencil mark until you get back to where you started. If you cut carefully, you will end up with one large circle.

MAGIC SQUARES

This square puzzle is very magical. When your friend arranges the pieces, she makes a square, but it has a hole in the middle. However, when you put the same puzzle together, the hole mysteriously disappears!

What You Need

An 8½ in × 11 in (21.5cm × 28cm) sheet of plain paper
A red and a black marker
Scissors
A ruler
A pencil

Preparation

1. Cut the piece of paper into a square. Measure *carefully* so that all sides are the same length.

2. Make a little pencil mark 3 in (7.5cm) to one side of each corner. Then connect the pencil marks as shown. Write out the words "MAGIC SQUARE" across each of the four sections with a red marker.

3. Erase the four little pencil marks and then *carefully* cut out the four sections.

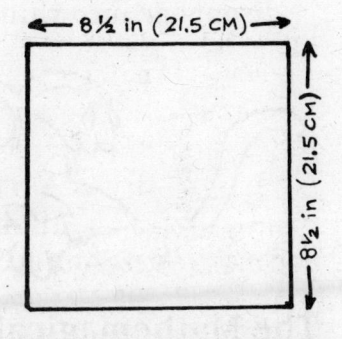

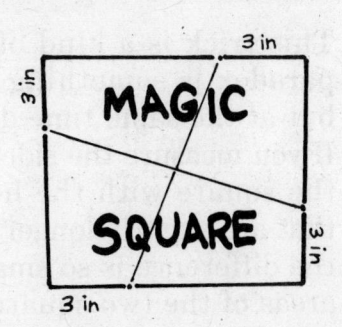

55

4. Turn the puzzle over and arrange the pieces as shown. Write the words "MAGIC SQUARE" across each of the four sections with a black marker.

What to Do

1. Place the four pieces with black letters faceup on the table and then mix them up. Ask your friend to put the puzzle together. When she is finished, her square will have a hole in the middle.

2. Turn the puzzle over and put together the side of the puzzle that has red letters. Your square will look like your friend's, but the hole will disappear into thin air!

The Mathemagical Secret

This trick is a kind of mathematical paradox. A paradox is something that seems to make sense but at the same time doesn't seem to make sense. If you measure the sides of each of the two puzzles, the square with the hole in the center has sides that are slightly longer (about 4% longer). Because the difference is so small, it's hard to tell that the areas of the two squares are different.

NUMBERLAND

Have you ever tried to fold a map? It's always difficult and you practically have to be a magician to fold it correctly. The Numberland map is a real challenge, and you will need all your magical skills to return it to its original shape.

What You Need

An 8½ in × 11 in (21.5cm × 28cm) sheet of plain paper   A pencil

Preparation

1. Divide a piece of paper into 8 equal sections by folding. Turn the paper sideways and then *carefully* fold it in half twice horizontally and once vertically.

2. Number each section as shown in the diagram.

3	4	2	7
6	5	1	8

front

57

3. Number the back of each section with the same number that appears on the front of the section.

7	2	4	3
8	1	5	6

back

What to Do

Fold the Numberland map so that only consecutively numbered sections are touching. For example, section 1 can only be touching section 2, section 4 can only be touching sections 3 and 5, and section 7 can only be touching sections 6 and 8. When you are finished, section 1 should be at the top and section 8 should be at the bottom. The answer appears on page 90.

The Mathemagical Secret

This puzzle uses a geometric procedure called a transformation. A transformation is a movement of shapes by flipping, turning, sliding, reflecting, or rotating.

Other Things to Do

Challenge your family and friends and see if they can fold the Numberland map.

58

5. Calculator Wizardry

Jungle Math

Think of a Number

Pumpkin Pi

Spooky

Final Score

Hide-and-Seek

Jungle Math

Here's a riddle for you. What might you get if you do your math homework in the jungle? If you don't know, just ask your calculator. It will tell you the answer!

What You Need

A calculator

What to Do

Ask your friend the Jungle Math riddle. If she doesn't know the answer, hand her a calculator and explain that it will solve the riddle for her. Then tell her to secretly:

1. Enter any number that is easy to remember. (This number must have less than 8 digits.)

   Example
   777

2. Double that number.

   $777 \times 2 = 1{,}554$

3. Subtract 58 from that answer.   1,554 − 58 = 1,496

4. Multiply that result by 3.   1,496 × 3 = 4,488

5. Divide that total by 6.   4,488 ÷ 6 = 748

6. Add 37 to that answer.   748 + 37 = 785

7. Subtract her original number from that result.   785 − 777 = <u>8</u>

What might you get if you do your math homework in the jungle?

You might get 8 (ate)!

The Mathemagical Secret

This trick was written using a kind of mathematics called algebra. Doubling and then multiplying by 3 is the same as multiplying by 6. Dividing by 6 cancels those operations. Every other operation is mathematical hocus-pocus that eliminates your friend's original number and guarantees the final total will always be 8.

THINK OF A NUMBER

Your friends will think that you possess supernatural powers when you mysteriously reveal their secret number!

What You Need

Paper and pencil   A calculator

What to Do

Hand your friend a calculator or have him work these problems using paper and pencil.

Tell him to secretly:	**Example**
1. Think of any number from 1 to 10.	5
2. Add 10 to that number.	5 + 10 = 15
3. Double that answer.	15 × 2 = 30
4. Subtract 4 from that total.	30 − 4 = 26
5. Double that answer again.	26 × 2 = 52
6. Divide that result by 4.	52 ÷ 4 = 13

Finally, remind your friend that he was free to choose any number. Then ask him to tell you his final total. Mentally subtract 8 from that total and your friend's secret number will magically appear!

$$13 - 8 = 5!$$

The Mathemagical Secret

This trick was written using algebra. Doubling twice and then dividing by 4 cancels those operations. The other operations add 8 to your friend's secret number. Subtracting 8 reveals that number.

Other Things to Do

1. Repeat the trick by having your friend choose another number from 1 to 10.

2. Have your friend enter a larger number into a calculator. It must be a number that is easy to remember and is less than 8 digits. At the end of the trick, have him hand you the calculator with his final total. Just subtract 8 to reveal his secret number.

Pumpkin Pi

Here is another riddle for you. What do you get when you divide the circumference of a pumpkin by its diameter? Pumpkin pi! In this trick, no matter what number your friend enters into a calculator, she always ends up with a piece of pi!

What You Need

A calculator

What to Do

Hand your friend a calculator and have her:

1. Secretly enter any number that is easy to remember. (This number must be less than 6 digits.)

   Example
   246

2. Add 12 to that number.

   $246 + 12 = 258$

3. Double that result.

   $258 \times 2 = 516$

4. Subtract 20 from that answer.

   $516 - 20 = 496$

5. Multiply that result by 25. $496 \times 25 = 12{,}400$

6. Add 57 to that answer. $12{,}400 + 57 = 12{,}457$

7. Divide that result by 50. $12{,}457 \div 50 = 249.14$

8. Subtract her original number. $249.14 - 246 = 3.14$

This trick can be repeated several times with the same friend. No matter what number she starts with, the final answer will always be 3.14. Since pi (π) = 3.1415926535 89793..., your friend will always end up with "a piece of pi!"

The Mathemagical Secret

This trick was written using algebra. Multiplying by 2 and then by 25 is the same as multiplying by 50. Dividing by 50 cancels those operations. Every other operation is mathematical hocus-pocus that eliminates your friend's original number and guarantees that the final total will always be 3.14!

Spooky

Your friend selects a card by cutting a deck of cards in half. After he works a few problems on a calculator, that card is turned over and it matches his final answer!

What You Need

A calculator   Paper and pencil
A deck of playing cards

Preparation

Put any 9-card on the *bottom* of the deck.

What to Do

1. Place the deck of cards on the table. Ask your friend to cut the deck by taking about half of the cards off of the top and putting them in a pile facedown on the table.

2. Complete the cut by taking the bottom half of the deck and placing it crosswise on top of his pile. The 9-card is now on the bottom of the top pile. This is called a cut force. After working the next part of the trick, your friend will forget the details of how the deck was cut.

Then ask your friend to:

1. Write any 7-digit number on a piece of paper. Tell him all seven digits must be different.

   Example
   3,906,275

2. Rearrange the seven digits in any order and write the new number below the first number.

   6,527,093

3. Subtract the smaller number from the larger number on a calculator.

   6,527,093
   −3,906,275
   2,620,818

4. Find the sum of the digits of his answer.

   2 + 6 + 2 + 0 + 8 + 1 + 8 = 27

If his answer has more than one digit, tell him to add those digits together until there is only one digit.

$$27 \rightarrow 2 + 7 = 9$$

Finally, remind your friend he was free to cut the deck wherever he wanted and free to choose any 7-digit number. Then ask him to turn over the top pile and look at the card that "he selected." It will be a 9-card and will match his final answer! Spooky!

The Mathemagical Secret

This trick uses a mathematical procedure called casting out nines, and it will work for any number of digits. The difference between the two numbers is always a multiple of 9, so the sum of its digits always equals 9.

Other Things to Do

Since this trick works for any number of digits, have your friend start with any phone number or the serial number from a dollar bill.

Final Score

Your friend writes down the final score of any sporting event, and you are able to divulge it by simply performing a little number magic on a calculator!

What You Need

A calculator   Paper and pencil

What to Do

Tell your friend to write secretly any final score from her favorite sport. Then hand her a calculator and ask her to:

Example
14 − 3

1. Enter the winning score without letting you see it.

    14

2. Multiply that number by 20.

    14 × 20 = 280

3. Add 5 to that total.   280 + 5 = 285

4. Multiply that answer by 50.   285 × 50 = 14,250

5. Add the losing score to that total.   14,250 + 3 = 14,253

6. Multiply that answer by 4.   14,253 × 4 = 57,012

7. Subtract 1,000 from that total.   57,012 − 1,000 = 56,012

Tell her to hand you the calculator with the final total. Just divide that total by 4,000 and your friend's final score will magically appear!

56,012 ÷ 4,000 = <u>14</u>. <u>003</u>

Winning Score **Losing Score**

Exceptions

If the final score is a tie, the winning and losing score is the same number.

Example: <u>4</u>. <u>004</u> so the final score is 4–4.

If you divide by 4,000 and there are not three digits after the decimal point, add zeroes so that there are three digits.

Examples
12.01 = <u>12</u>. <u>010</u>
so the final score is 12–10.

109.1 = <u>109</u>. <u>100</u>
so the final score is 109–100.

If you divide by 4,000 and there are no numbers after the decimal point, the losing score is 0.

Examples
3 = <u>3</u>. <u>000</u> so the final score is 3–0.
0 = <u>0</u>. <u>000</u> so the final score is 0–0.

The Mathemagical Secret

This trick was written using algebra. Multiplying by 20 and then by 50 is just like multiplying by 1,000. This moves the winning score over to the left of the hundreds place. Adding the losing score puts that number in the last three places. Dividing by 4,000 puts the winning score on the left side of the decimal point and the losing score on the right side.

Hide-and-Seek

Your friend secretly works a subtraction problem and then tries to conceal one of the digits of his answer. Within seconds, you are able to reveal the missing digit!

What You Need

A calculator   Paper and pencil

What to Do

1. Have your friend write down any 3-digit number without letting you see it. Tell him that all three digits must be different.

   Example
   394

2. Tell him to rearrange the three digits in any order and write this new number next to his first number.

   394   943

3. Ask him to subtract the smaller number from the larger number.

   943
   −394
   549

4. Tell him to circle one digit in his answer that is *not* a 0.

   ⑤49

5. Ask him to read off the remaining digits in any order.

   9, 4

6. Mentally add the digits he reads to you.

   9 + 4 = 13

If the answer has more than one digit, add those digits together until only one is left.

$13 \to 1 + 3 = 4$

7. Mentally subtract that number from 9 and the missing digit is magically revealed!

$9 - 4 = 5$

Exceptions

Example

1. If the sum of the remaining digits is 9, the missing digit is 9.

```
  674
 -476
  1⑨8
```

2. When your friend ends up with a 2-digit number, he will read only one digit. Just subtract it from 9 to get the missing digit.

```
  561
 -516
   ④5
```

$9 - 5 = 4$

3. When your friend ends up with a 1-digit number, he won't read any digits. The missing digit will be 9.

```
  965
 -956
    ⑨
```

The Mathemagical Secret

This trick uses a mathematical procedure called casting out nines, and it will work for any number of digits. The difference between the two numbers is always a multiple of 9, so the sum of its digits always equals 9.

6. Mixed Bag of Tricks

Tight Fit

9-Tailed Cat

Impossible Will

Vegas

Handcuffs

Incredible Memory

TIGHT FIT

10,000,000,000,000,000,000,000,000,000,000,
000,000,000,000,000,000,000,000,000,000,000,
000,000,000,000,000,000,000,000,000,000,000

It would be an incredible magic trick if you could enter this entire number into your calculator's display. Actually, it is possible. Carefully work this math problem and then turn your calculator upside down.

99,999,999 ÷ 9 − 11,058,162 + 656,060 =

The answer is GOOGOL. A googol is one of the largest numbers that has a name. It is 1 followed by 100 zeroes, and when written out, looks like the number above.

Do you think that googol is a strange name for a number? Many years ago, an American mathematician by the name of Edward Kasner thought 1 followed by 100 zeroes should have a name. So he asked his 9-year-old nephew, Milton Sirotta, to name it. The first word that came out of Milton's mouth was "googol," and it has been called that ever since.

How Large Is a Googol?

It is more than the number of grains of sand in the entire world. It has been estimated that the number of grains of sand needed to fill a sphere the size of the earth would be 1 followed by 32 zeros. It is even more than the number of elementary particles in the entire universe. All of the protons, neutrons, and electrons add up to 1 followed by 80 zeroes.

Googolplex

A googolplex is a much larger number than a googol. A googolplex is 1 followed by a googol of zeroes!

How Large Is a Googolplex?

It is an unbelievably large number! It is so large that you could spend your entire life counting to it, and you would never get there. It is also impossible to write a googolplex. You could not write it even if you traveled to the farthest star and back and wrote a zero every inch of the way!

9-Tailed Cat

You would be one of the greatest magicians of all time if you could prove a cat has nine tails. Here is some mathematical trickery that will have you wondering if it's true!

Proof:

1. Have you ever seen a cat that has 8 tails? Have you ever talked to anyone who has seen a cat that has 8 tails? Neither have I. Therefore, this statement must be true.

    No cat has 8 tails.

2. If you enter a room and there are no cats, you see 0 tails. But, if you enter a room and there is a cat, you see 1 tail. Therefore, this statement must be true:

    A cat has one more tail than no cat.

3. So, if no cat has 8 tails *and* a cat has 1 more tail than no cat (8 + 1 = 9), then this statement must be true:

    A cat has 9 tails!!!

The Mathemagical Secret

In Step 1 "no cat has 8 tails" means that "there aren't any cats that have 8 tails." It does not mean that "zero cats have 8 tails." Therefore, the mathematical reasoning is incorrect in Step 2 where "no cat" means "zero cats."

IMPOSSIBLE WILL

A man died and left 17 valuable coins to his three children. However, it's impossible to divide the coins as it is written in the will unless you know the mathemagical secret!

What You Need

18 coins   Paper and pencil

Preparation

Put one coin in your pocket. Copy the Last Will and Testament onto a piece of paper.

LAST WILL AND TESTAMENT
I leave my 17 valuable coins to my three children. I would like the coins divided in the following manner:

$1/2$ of the coins to my oldest child.
$1/3$ of the coins to my middle child.
$1/9$ of the coins to my youngest child.

What to Do

1. Show your friend the dead man's Last Will and Testament. Tell him there was a big problem when all the family members gathered for the reading of the will. The judge said it was completely legal, but he did not know how to divide the coins among the children as requested in the will.

2. Put any 17 coins on the table, and then ask your friend if he can divide them as it is stated in the will. Of course he won't be able to do this because 17 is not divisible by 2, 3, or 9.

3. Show your friend how to solve the problem. Explain that the dead man's lawyer happened to have a valuable coin in her pocket. She took out her coin and added it to the pile of 17 coins to make a total of 18 coins.

(Remove the coin from your pocket and add it to the other coins.)

Then she divided the coins as follows:

Oldest child—$1/2$ of 18 coins = 9 coins

(Put 9 coins into a pile.)

Middle child—$1/3$ of 18 coins = 6 coins

(Put 6 coins into a second pile.)

Youngest child—$1/9$ of 18 coins = <u>2 coins</u>

(Put 2 coins into a third pile.)

Total = 17 coins

The total was only 17 coins, so there was one coin left over that the lawyer put back in her pocket!

(Put the last coin back in your pocket.)

Each of the children received a fair share of coins and the lawyer got her coin back!

The Mathemagical Secret

The reason why the judge could not divide the coins properly is because 17 is not divisible by 2, 3, or 9, and because $1/2 + 1/3 + 1/9$ does not equal 1.

$$\begin{aligned} 1/2 &= 9/18 \\ 1/3 &= 6/18 \\ + 1/9 &= 2/18 \\ \hline \text{Total} &= 17/18 \end{aligned}$$

So one more coin must be added to the pile of 17 coins to make a total of 18/18, or one whole.

$$17/18 + 1/18 = 18/18 \text{ or } 1.$$

Vegas

Your friends are amazed that you are able to reveal the sum of the numbers on three dice even though the dice are rolled when your back is turned!

What You Need

3 dice ("Monopoly," "Yahtzee," and many other board games have dice.)

What to Do

While your back is turned, have a friend:

1. Roll three dice.

2. Find the sum of the three top numbers.

3. Pick up one of the three dice and add its bottom number to his total.

4. Roll that die one more time and add its top number to his total.

When your friend is finished, turn around and pretend to perform some hocus-pocus as you look over the dice. Even though you have no idea which die was rolled a second time, you can tell your friend's total by just adding 7 to the sum of the top numbers that are showing on the three dice! So in the example, your friend's total is 13 + 7, or 20!

The Mathemagical Secret

On any die, the sum of the top number and the bottom number is always 7. Adding the bottom number and then rolling that same die again adds 7 to the sum of the three top numbers that are showing.

Other Things to Do

1. Use more dice if you repeat the trick. Your friend's total will still equal the sum of the top numbers on the dice plus 7 no matter how many dice you use.

2. Let your friend pick up two dice in Step 3 and then roll both of them in Step 4. Then his total will equal the sum of the top numbers on the dice plus 14.

Handcuffs

"Those cuffs can be tricky!..."

This is a magic trick that seems impossible and will leave everyone completely baffled! Even though a string is securely tied to each of your wrists, you are able to attach a ring to the middle of the string!

What You Need

A 50 in (127cm) piece of string   A ring or bracelet

What to Do

1. Have your friend tie your wrists together with a string (Figure A). Have her tie it tight enough so that you can't slip the string over your hands, but not so tight that it cuts into your wrists.

A

2. Explain to your friend that you are going to attach a ring to the middle of the string without removing the string from your wrists.

3. Take the ring and then turn around so your friend can't see what you are doing.

4. Make a large loop with the center of the string and then push the loop through the center of the ring (Figure B).

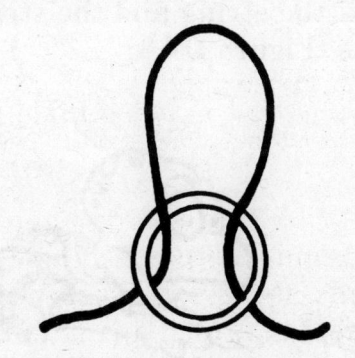

B

5. Push the large loop under the string that is tied around your left wrist. Then loop it up and over the top of your left hand (Figure C).

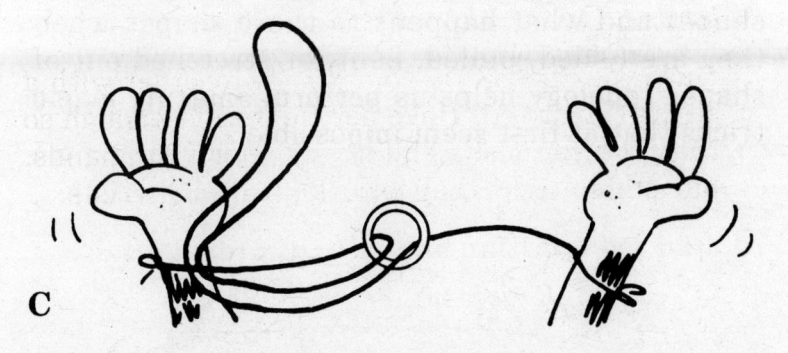

C

6. Slip the loop that is not tied to your wrist up and over the top of your left hand.

85

D

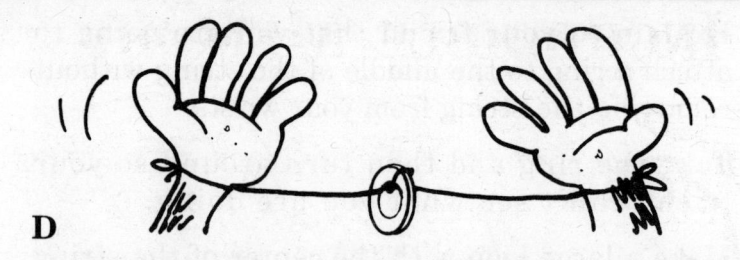

7. That's it. Turn around and show your friend. She won't believe her eyes. The ring is attached to the center of the string and the string never left your wrists (Figure D)!

The Mathemagical Secret

This is another trick that uses the kind of mathematics called topology. Topology is the study of shapes and what happens to those shapes when they are folded, pulled, bent, or stretched out of shape. Topology helps us perform amazing magic tricks that at first seem impossible.

INCREDIBLE MEMORY

Your friends are astonished when you show them you have memorized nine 9-digit numbers!

What You Need

9 index cards   Paper and pencil

Preparation

Copy these numbers onto index cards, one to each card. Circle the card numbers.

1. 774,156,178
2. 415,617,853
3. 123,583,145
4. 820,224,606
5. 538,190,998
6. 246,066,280
7. 943,707,741
8. 651,673,033
9. 369,549,325

Example:

⑧
651,673,033

What to Do

1. Mix up the index cards so they are not in order and hand them to your friend.

2. Tell him there is a different 9-digit number written on each card and you have memorized all of the numbers. (You don't have to memorize any numbers. You just multiply to get the first two digits and then add to get the remaining seven digits.)

3. Ask him to pick any card and tell you the card number.

4. When your friend tells you the card number, mentally multiply it by 7 and then reverse your answer. The result is the first two digits of the 9-digit number.

Example: Card #8
8 × 7 = 56 and 56 reversed is 65

So write down the first two digits of the number.   65

5. To get the next digit, mentally find the sum of the first two digits. If this sum is less than 10, write it down. If it is 10 or greater, only write down the number that is in the ones place.

6 + 5 = 1<u>1</u>   65<u>1</u>

6. Continue adding the last two digits to get the next digit until you have written down all nine digits.

5 + 1 = <u>6</u>   651,<u>6</u>
1 + 6 = <u>7</u>   651,6<u>7</u>
6 + 7 = 1<u>3</u>   651,67<u>3</u>
7 + 3 = 1<u>0</u>   651,673,<u>0</u>
3 + 0 = <u>3</u>   651,673,0<u>3</u>
0 + 3 = <u>3</u>   651,673,03<u>3</u>

The number on card #8 is 651,673,033.

An Exception

If your friend picks card #1, 1 × 7 = 7, which is only one digit. So just repeat the digit 7 to get the first two digits. The number on card #1 is 774,156,178.

Other Things to Do

If you really want to impress your friends, prepare cards that have more than 9 digits.

Answer to "Numberland" (Pages 57–58)

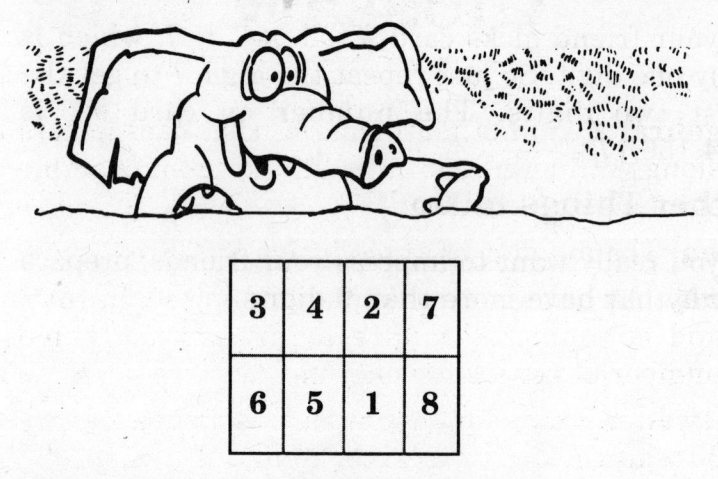

3	4	2	7
6	5	1	8

Fold sections 3 and 6 back and under.

4	2	7
5	1	8

Fold sections 5, 1, and 8 up and over. Then turn the paper around so that you can read the numbers.

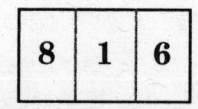

8	1	6

Fold section 6 back and under and then fold section 8 back and under.

1

Glossary

algebra A kind of mathematics that uses letters along with numbers. $7x + 2 = 23$ is an example of an algebra problem.

area The amount of space inside a figure.

array An orderly arrangement of objects in rows and columns. A square array has the same number of rows and columns.

between Example: A number is between 20 and 30 when it is greater than 20 and less than 30.

cancel One operation eliminates another operation. Example: Subtracting 7 cancels adding 7.

casting out nines A process that removes groups of nine from a number. Example: $23 - 9 = 14$ and $14 - 9 = 5$. Quick way: $23 \rightarrow 2 + 3 = 5$.

circumference The distance around a circle.

column Objects that are arranged vertically.

consecutive numbers Numbers that are in order. Example: 3, 4, 5, and 6.

cut force A way of cutting a deck of cards that forces someone to choose a certain card.

decimal point A period used in decimal numbers to separate the whole number part from the decimal part.

diagonal, diagonally Running in a slanted direction.

diameter The distance across the center of a circle.

die (plural: **dice**) One of a set of dice.

difference The answer to a subtraction problem.

digit Any of the symbols 0 through 9 used to write numbers. Example: 358 is a 3-digit number.

divisible Can be divided with a remainder of zero. Example: 18 is divisible by 2 because 18 ÷ 2 = 9.

double Multiply by 2.

edge of a piece of paper The line segment where the front side and the back side meet.

elementary particles Protons, neutrons, and electrons.

ESP The ability to read minds, sense distant happenings, and/or predict the future.

even numbers The numbers 0, 2, 4, 6, 8, 10, . . .

face A flat surface of a solid figure. Example: The flat surface of a die.

face cards The Jacks, Queens, and Kings of playing cards.

from Example: A number from 1 to 10 includes the numbers 1 through 10.

googol 1 followed by 100 zeroes.

googolplex 1 followed by a googol of zeroes.

hocus-pocus Extra steps that are added to a magic trick that help hide the trick's secret. Extra showmanship that makes a trick more interesting.

horizontal, horizontally Running from left to right.

hundreds place Example: 2,538. The 5 is in the hundreds place.

infinity A number that has no limits.

magic square An array of numbers where the sum of each row, column, and diagonal is the same.

mathemagic Magic tricks that use numbers.

millennium A period of 1,000 years.

multiple Example: The multiples of 9 are: 9 × 0 = **0**, 9 × 1 = **9**, 9 × 2 = **18**, 9 × 3 = **27**, etc.

multiplication table A chart used for learning multiplication facts.

odd numbers The numbers 1, 3, 5, 7, 9, . . .

ones place Example: 2,538. The 8 is in the ones place.

operations add (+), subtract (−), multiply (×), divide (÷), raise to a power (^{x}), and take a square root ($\sqrt{\ }$).

paradox Something that seems to make sense but at the same time doesn't seem to make sense.

parallel lines Lines in a plane that stay exactly the same distance apart.

parallelogram A quadrilateral whose opposite sides are parallel and equal.

pattern A set of numbers arranged in a certain order. Example: 1, 2, 3 or 8, 6, 4.

pi (π) The number obtained by dividing the circumference of a circle by its diameter. It is approximately equal to 3.14.

plane A flat surface that extends without end in all directions.

polygon A closed 2-dimensional figure with three or more sides.

product The answer to a multiplication problem.

proof The process that establishes that a statement is true.

psychic Sensitive to supernatural forces.

quadrilateral A polygon with four sides.

quotient The answer to a division problem.

random number A number chosen by chance.

rectangle A polygon with four right angles.

right angle An angle that measures 90°.

row Objects that are arranged horizontally.

secret card The chosen card in a card trick.

sphere A ball-shaped 3-dimensional figure.

square A polygon with four right angles and four equal sides.

square a number Multiply a number by itself. Example: 9 squared = 9 × 9 = 81.

sum The answer to an addition problem.

tens place Example: 2,538. The 3 is in the tens place.

thousands place Example: 2,538. The 2 is in thousands place.

topology A kind of mathematics that studies shapes and what happens when they are folded, pulled, bent, or stretched out of shape.

transformation A movement of shapes by flipping, turning, sliding, reflecting, or rotating.

vertical, vertically Rising up and down.

ABOUT THE AUTHOR

Raymond Blum has been a mathematics teacher for over 30 years. In 1991, he wrote the book *Mathemagic*, which is filled with dozens of entertaining number-magic tricks. His second book, *Math Tricks, Puzzles & Games*, was published in 1994. It is a fascinating collection of fun-filled math activities. In 1997, he wrote his third book, *Mathamusements*, which contains exciting mathematical fun from A to Z. All three books are written for children ages nine and up and for teachers to use in their classrooms.

Ray Blum has been a speaker at numerous state and national math conferences, where he shares his classroom magic with other teachers. For years, he has performed a number-magic show for elementary and middle school children as "Professor Numbers." The professor shows children the magical, fun side of mathematics with his mathemagic and arithmetricks.

Ray has won several awards for his teaching, including 1994 Wisconsin Teacher of the Year, and is currently teaching eighth grade math at Spring Harbor Middle School in Madison, Wisconsin.

ABOUT THE ILLUSTRATOR

Jeff Sinclair has been drawing cartoons ever since he could hold a pen. He has won several local and national awards for cartooning and humorous illustration. When he is not working away at his drawing board, Jeff can be found renovating his home and feeding hungry koi in his backyard water garden. Jeff has recently gone into cyberspace on the Internet. He currently lives in Vancouver, BC, Canada, with his wife, Karen; son, Brennan; daughter, Conner; and golden lab, Molly.

Index

Algebra, 61, 63, 65, 71
Banana trick, 16–18
Bashful King, 35–36
Calculator tricks, 59–73
Cards, tricks with, 21–36
Casting out nines, 68, 73
David Bananafield, 16–18
Dice, 8–9, 82–83
Dictionary, 12–13
Easy Money, 30–32
ESP, 40–42
Final Score, 69–71
Googol, 75–76
Googolplex, 76
Handcuffs, 84–86
Hide-and-Seek, 72–73
Impossible Will, 79–81
Incredible Memory, 87–89
Infinity, 53–54
Jungle Math, 60–61
Kasner, Edward, 75
King, Bashful, 35–36

Last Will and Testament, 79–81
Lucky Joker, 27–29
Magic Spell, 22–23
Magic Squares, 55–56
Map, folding a, 57–58
Mastermind, 38–39
Math Wizard, 19–20
Memory, Incredible, 87–89
Millennium, 50–51
Mind-Boggling, 48–49
Mindreading tricks, 7–20
Mixed Bag of Tricks, 74–89
Möbius Strip, 54
Multiplication Table, 14–15
Nines, casting out, 68, 73
Nine-Tailed Cat, 77–78
Numberland, 57–58; answer to, 90
Odds and evens, 27–29, 35–36

Paper Magic, 52–58
Paradox, 56
Perfect Match, 45–47
Pi, pumpkin, 64–65
Psychic Predictions, 37–51
Pumpkin Pi, 64–65
Ring trick, 84–86
Secret Word, 12–13
Sirotta, Milton, 75
Skyscraper, 8–9
Spooky, 66–68
Sporting event, 69
String trick, 84–86
Think of a Number, 62–63
Three Strikes You're Out!, 33–34
Thumbprint, 10–11
Tight Fit, 75–76
Topology, 54, 86
Topsy-Turvy, 43–44
Transformation, 58
Vegas, 82–83
Whispering Deck, 24–26